Nature's Weapons

How Her Creatures Defend Themselves

By Jeremy Staines

First Edition February 2017

Published by Wild Creatures Press

An interesting part of the animal world is the knowledge which the members possess of the surest and quickest way to kill their prey or enemies. Many living things must sustain their life by eating some other living thing. That seems a very cruel and unfair arrangement, especially for the weaker animals, but in nearly every case of killing the stronger animal knows the exact spot to strike to cause immediate death.

A dog or ferret knows exactly how to inflict instantaneous death upon a rabbit or a hare; an eagle or a raven goes as surely to the right spot for attack upon a fallen deer or weakened sheep as a fox goes to the fatal spot in a domestic fowl. And a humble spider, spinning its silken web in our garden, is as unerring in their death stroke as a cheetah that fastens its teeth in the straining throat of a buck.

Whether the wound is made with sledge hammer paws and vice-like jaws, or with the poison fang and stiletto of the insect, the end is quick.

You know how fierce a rat is and how hard are its teeth and strong its jaws, yet an owl will strike it dead, and gobble it up in a twinkling. When cornered a rat will fly at a human being, yet against this bird it is always putting up a losing fight.

Birds are unerring warriors. The beautiful heron will spear an enemy or its prey with deadly accuracy.

Some of nature's creatures use poison to help them secure their food, and as a means of defence. The cobra eats the same meat food as a lion. When in need of food it aims at a victim, and bites, perhaps making a wound no bitter than a pinprick. But as the fang punctures, a squirt of virus enters the wound. The creature leaps on its way, but its very pace makes death the speedier. The blood infected by the poison will reach the heart in a few seconds, and it drops dead. The effect is almost as quick as a bullet.

Reptiles also have a weak point, and the mongoose and kookaburra know it. The kookaburra dives on the snake, holds it near the head, and flying to a high tree, lets it fall and stuns it.

The little furry mongoose is as wary in the combat as a swordsman in a duel; it feints and dodges, it rushes and eludes, watching for its chance to bite behind the neck, and fatally injure the snake. A poisonous snake with venom enough to kill a houseful of people is breakfast, dinner and tea to his majesty the mongoose. Teeth, speed and courage defeat venom, fang and fury.

You would think that the crocodile at the Zoo was all right at defence with his hard shell, but very harmless at attacking. But not one visitor in a thousand realises that in that mail-covered, motionless log there is one of the most formidable fighting machines in the world. It is not in those grim jaws, but in that tremendous tail. The tail is all muscle and bone, and it can be swung rapidly to and fro like a springy cane. It can send the biggest man sprawling; it can knock a large animal off its feet, and if the first blow misses, back it comes like lightning. Then comes death by drowning.

Have you ever seen a wasp and a spider battle? It is a demonstration of skill between two of the most finished warriors in the world. If the wasp gets into the web she is a spider's meal, but if the spider is out of the web it is the wasp's turn.

She can give the spider a thrust with her sting, and the spider makes the meal. Wasps and spiders are more astonishing than the most poisonous snake. If a venomous snake bites an animal, the victim must die. It is not so with the quarry of spider and wasp. A spider can bite – poison or not – as she desires. Of all the multitudes of different spiders, everyone has fangs and poison. It seems clearly that if the bite is a deemed sufficient the spider will save her poison, but if the grip of the fangs has not ended the struggles poison is poured instantly into the victim.

A wasp also exercises discretion in battle with a spider. If a meal is required urgently the wasp stings the spider in what we may call the neck. That spells immediate death. But the spider may be wanted for food weeks hence for a wasp yet unborn – a wasp that will emerge from an egg that the hunting mother is about to lay. If the spider were killed at once, its body would go bad before the baby wasp came out of the egg. So, with a wisdom that passes our understanding, the wasp stabs the spider through the chest, placing its poison in an important nerve centre, so that the spider becomes paralysed. It is insensible to suffering, yet continues to live until Nature's appointed hour, when the baby wasp will be hatched and hungry. The paralysed spider is carried in this condition to the nest, and there imprisoned. A sting which can kill or permit life to linger for weeks at once.

A queen bee knows how to strike home as surely as a sword fish knows how to spear a whale. It is one of the miracles of Nature that two queen bees in combat in a hive, if they reach a position in which a simultaneous thrust of stings will bring death to both, withdraw and alter their pose, so that, while death must come one shall be left to carry on the duty of peopling the hive.

These are but a few examples of animal efficiently in the use of weapons. The subject is perhaps a little terrifying, but throughout we see clearly that, fearful as the daily battle in Nature really is, there is sudden onslaught and a quick end to suffering.

It is as if Nature, resolved upon having no favourites, giving all a chance to live and desired to be merciful in death. Better for a wild animal to meet a sudden violent end than to wax old and feeble and perish of gnawing disease and creeping starvation.

THE END

BIBLIOGRAPHY

1931 'Nature's Weapons--How Her Creatures Defend Themselves.', *The Age (Melbourne, Vic. : 1854 - 1954)*, 29 May, p. 1. (Supplement FOR YOUNG PEOPLE), viewed 14 Feb 2017, http://nla.gov.au/nla.news-article205841540